MW01388875

Our Exciting Earth!
OCEANS

By Claire Romaine

Gareth Stevens
PUBLISHING

Please visit our website, www.garethstevens.com. For a free color catalog of all our high-quality books, call toll free 1-800-542-2595 or fax 1-877-542-2596.

Cataloging-in-Publication Data

Names: Romaine, Claire.
Title: Oceans / Claire Romaine.
Description: New York : Gareth Stevens Publishing, 2018. | Series: Our exciting Earth! | Includes index.
Identifiers: ISBN 9781538209691 (pbk.) | ISBN 9781538209714 (library bound) | ISBN 9781538209707 (6 pack)
Subjects: LCSH: Ocean–Juvenile literature. | Oceanography–Juvenile literature.
Classification: LCC GC21.5 R65 2018 | DDC 551.46–dc23

Published in 2018 by
Gareth Stevens Publishing
111 East 14th Street, Suite 349
New York, NY 10003

Copyright © 2018 Gareth Stevens Publishing

Editor: Therese Shea
Designer: Bethany Perl

Photo credits: Cover, p. 1 Galyna Andrushko/Shutterstock.com; p. 5 Veronika Vankova/Shutterstock.com; p. 7 Melanie Hobson/Shutterstock.com; p. 9 Tischenko Irina/Shutterstock.com; p. 11 Franco Banfi/WaterFrame/Getty Images; pp. 13, 24 SAPWUT KUNDEJ/Shutterstock.com; pp. 15, 24 stephan kerkhofs/Shuttestock.com; pp. 17, 24 Laura Dinraths/Shutterstock.com; p. 19 Jurjen Veerman/Shutterstock.com; pp. 21, 24 Aleksey Stemmer/Shutterstock.com; p. 23 Alexey Y. Petrov/Shutterstock.com.

All rights reserved. No part of this book may be reproduced in any form without permission in writing from the publisher, except by a reviewer.

Printed in the United States of America

CPSIA compliance information: Batch #CW18GS: For further information contact Gareth Stevens, New York, New York at 1-800-542-2595.

Contents

Visit the Ocean! 4

Ocean Animals 8

Ocean Plants 14

Tides 18

Let's Swim! 22

Words to Know 24

Index 24

Let's go to the ocean!

5

Oceans are big.
They contain salt water.

7

The ocean has many animals.
Fish are just one kind!

9

The biggest
ocean animal is
the blue whale!

11

Corals are
ocean animals.
They look like plants!

13

Ocean animals need ocean plants.
Sea turtles eat them!

15

Sea grass is an ocean plant.
Sea horses hide in it!

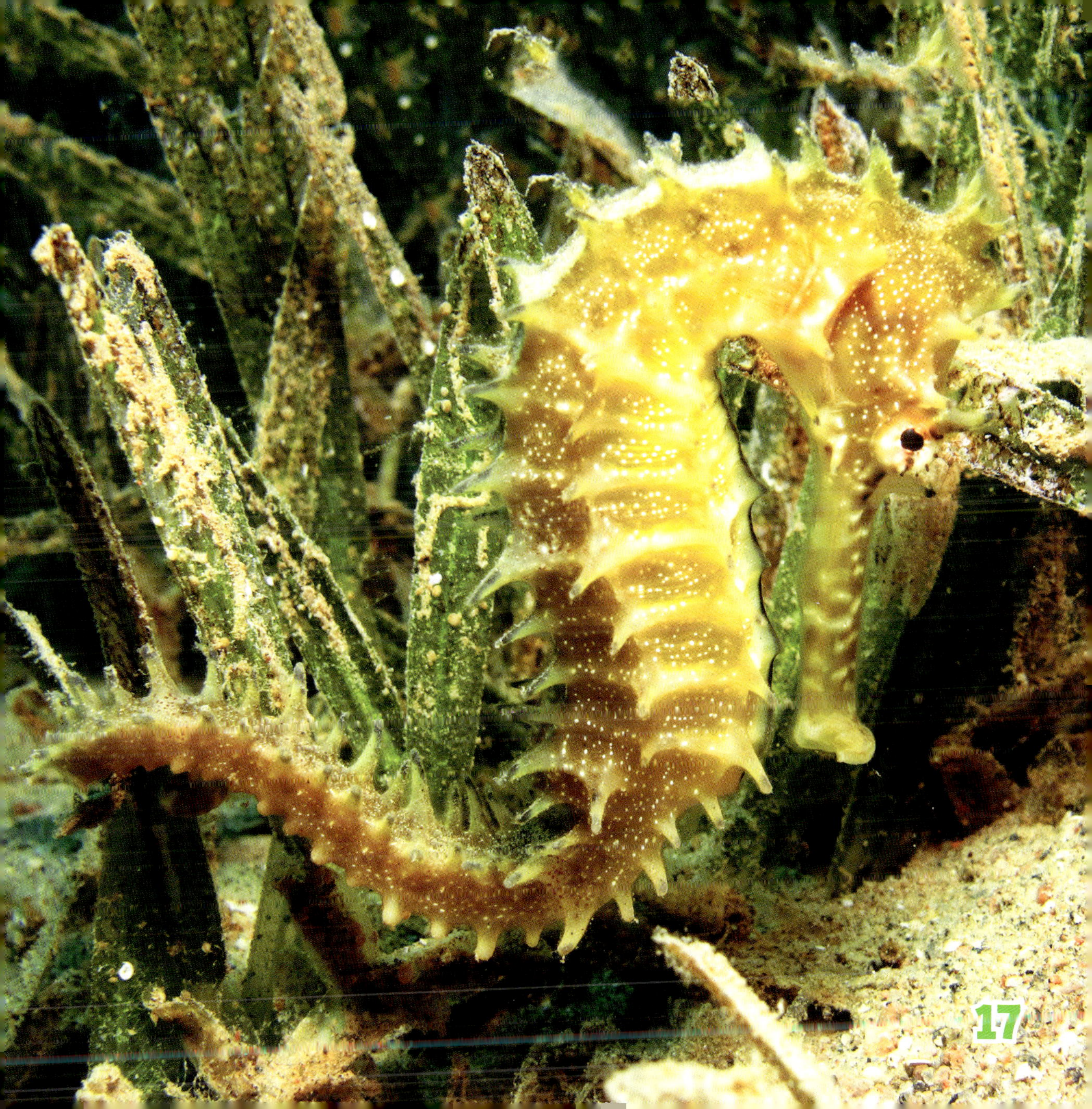

Ocean water rises and falls.
This is a tide.
The moon and sun cause tides!

19

We look for
ocean animals.
I see a crab!

21

We swim in the ocean.
Oceans are fun!

23

Words to Know

corals   crab   sea horse   sea turtle

Index

blue whale 10
corals 12
crabs 20

sea horses 16
sea turtles 14
tide 18

24